SpringerBriefs in Water Science and Technology

More information about this series at http://www.springer.com/series/11214

William F. Hunt · Bill Lord
Benjamin Loh · Angelia Sia

Plant Selection
for Bioretention Systems
and Stormwater Treatment
Practices

William F. Hunt
North Carolina State University
Raleigh, NC
USA

Bill Lord
North Carolina State University
Louisburg, NC
USA

Benjamin Loh
Baxter Design Group
Queenstown
New Zealand

Angelia Sia
National Parks Board
Singapore
Singapore

Prof. William F. Hunt is Research Fellow, Centre for Urban Greenery & Ecology (2012)
The CUGE Research Fellowship is a scheme that offers international researchers an opportunity to conduct and lead research relevant to the urban greenery and ecology of Singapore. Details are found in https://www.cuge.com.sg/research/CUGE-Research-Fellowship

ISSN 2194-7244   ISSN 2194-7252 (electronic)
ISBN 978-981-287-244-9   ISBN 978-981-287-245-6   (eBook)
DOI 10.1007/978-981-287-245-6

Library of Congress Control Number: 2014951686

Springer Singapore Heidelberg New York Dordrecht London

Printed on acid-free paper

Springer is part of Springer Science+Business Media (www.springer.com)

Acknowledgments

We thank the National Parks Board, Singapore Delft Water Alliance—National University of Singapore, Public Utilities Board, and Dr. Tan Puay Yok for their support in the project "Selection of Plants for Bioretention Systems in Singapore."

Acknowledgements

Contents

Contents

Chapter 1
Introduction

Bioretention systems, also known as biofiltration systems, biofilter or rain gardens, is a common stormwater mitigation measure. It utilises a low energy consumption treatment technology to increase water quality and reduce peak discharge.

A typical bioretention system can be configured as a basin or a longer narrower vegetated swale overlaying a porous filter medium with a drainage pipe at the bottom. Surface runoff is diverted from the kerb or pipe into the biofiltration system, where it physically filtered through dense vegetation and temporarily ponds on the surface of planting media that acts as a filter before being slowly infiltrated vertically downwards through the media. Depending on the design, treated water—the effluents—is either exfiltrated into the underlying or surrounding soils, or collected in the underdrain system—subsoil perforated drain—to downstream waterways or receiving waterbodies. The system varies in size and receives and treats runoff from a variety of drainage areas within a land development site. They can be installed in parks, roadside planting verges, parking lot islands, commercial areas, civic squares and other unused areas.

© The Author(s) 2015
W.F. Hunt et al., *Plant Selection for Bioretention Systems and Stormwater Treatment Practices*, SpringerBriefs in Water Science and Technology, DOI 10.1007/978-981-287-245-6_1

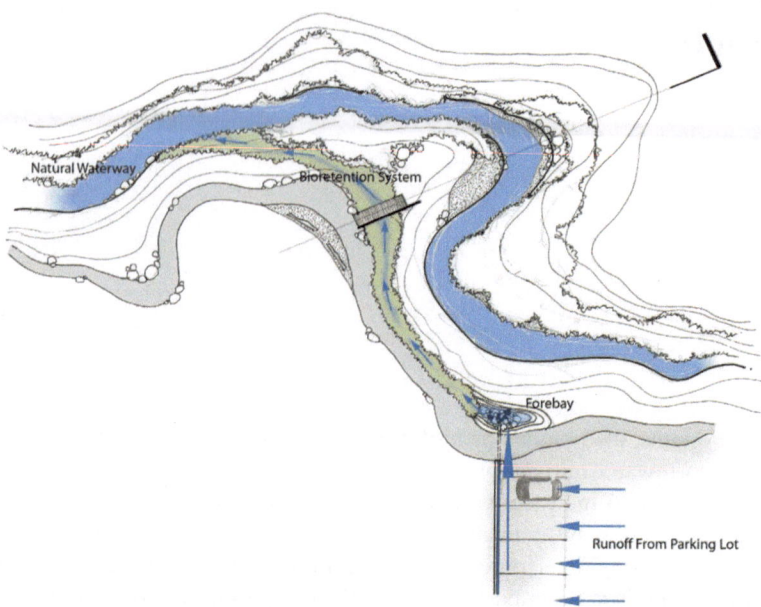

Runoff from the impervious surfaces such as carpark area can be diverted into drain and channelledinto bioretention system for treatment before it is being discharged into the receiving waterway

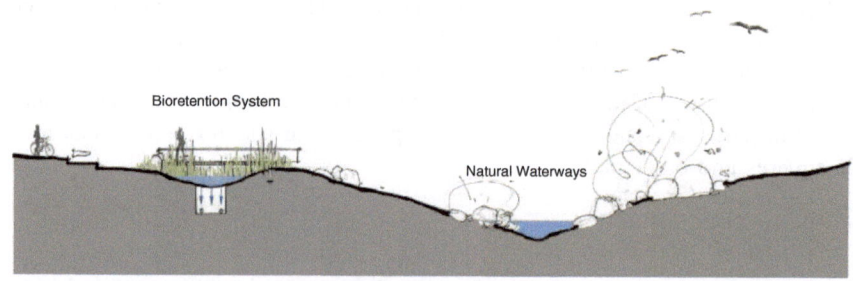

Cross section view of the parkland with a bioretention system integrated into the design of the space

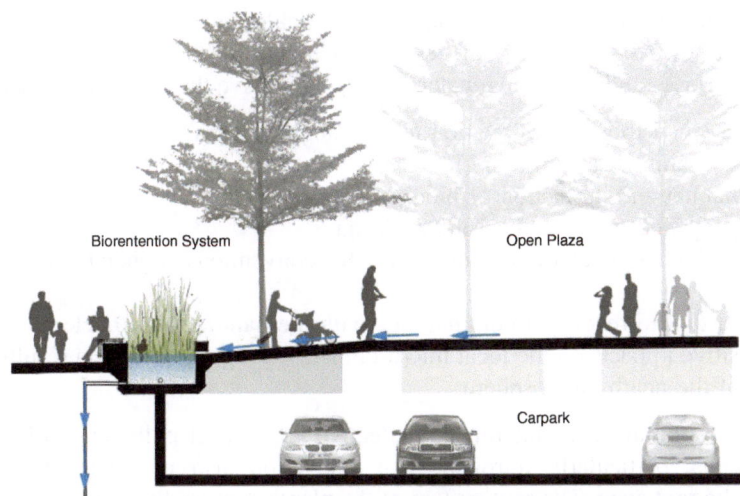

Bioretention system can also be designed and constructed above grade level. In this illustration bioretention systems designed above a carpark to treat stormwater as well as to create a buffer between the open space and main pedestrian circulation path

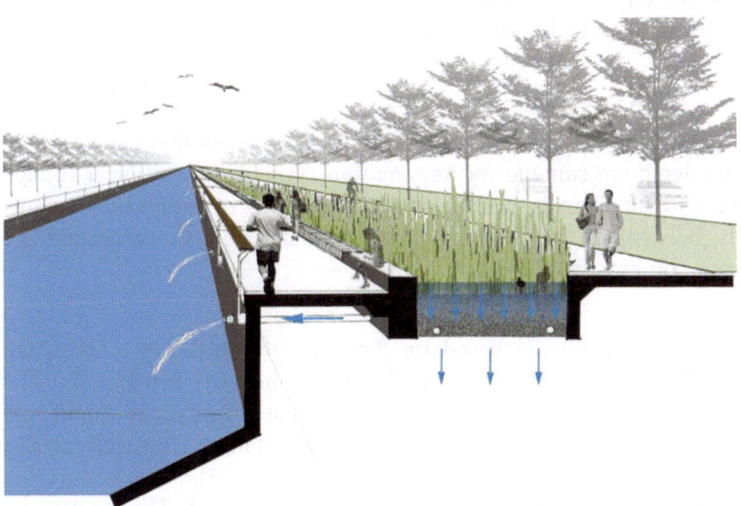

Bioretention system can be designed to provide visual as well as ecological connectivity within strategic open space network

Bioretention systems have been found to be viable and sustainable as water treatment devices. In addition to the ability to reduce peak flow generated by impervious surfaces and improving water quality, they have the following benefits:

- Take up a small footprint in relation to its catchment area
- Are self-irrigating (and fertilizing)
- Provide habitat and protection of biodiversity
- Can be integrated with the local urban design
- Assume a higher level of amenity than the conventional concrete drainage system
- Serve as a tool to reconnect communities with the natural water cycle
- Have positive impacts on the local micro-climate—evapotranspiration results in cooling of the nearby atmosphere

Plants are essential for facilitating the effective removal of pollutants in bioretention systems, particularly nitrogen. The vegetation also maintains the soil structure of the root zone. The root system of the plants continually loosens the soil and creates macropores, which maintain the long-term infiltration capacity of bioretention systems. Some plant species are more effective than others in their ability to adapt to the conditions within a biofilter.

The key parameters to consider for selecting plant types for bioretention systems are:

Growth form

Plant species that have extensive root structure with deep roots that penetrate the entire filter media depth are suitable for bioretention systems. Dense linear foliage with a spreading growth form is desirable, while bulbous or bulbo-tuber plants should generally be avoided as they can promote preferential flows around the clumps, leading to soil erosion.

Water requirement

Plant material selection should be based on the goal of simulating a terrestrial vegetated community which consists of shrubs and groundcovers materials. The intent is to establish a diverse, dense plant cover to treat storm water runoff and withstand urban stresses from insect and disease infestations, as well as the hydrologic dynamics of the system.

There are essentially three zones within a bioretention system. The lowest elevation supports plant species that are adapted to standing and fluctuating water levels. The middle elevation supports a slightly drier group of plants that grows on normal planting media, but with some tolerance to fluctuating water levels. The outer edge is the highest elevation and generally supports plants adapted to dryer conditions as it is above the ponding level.

"Wet footed" plants, that is obligate wetland species, are generally not recommended if the filter media used is sandy.

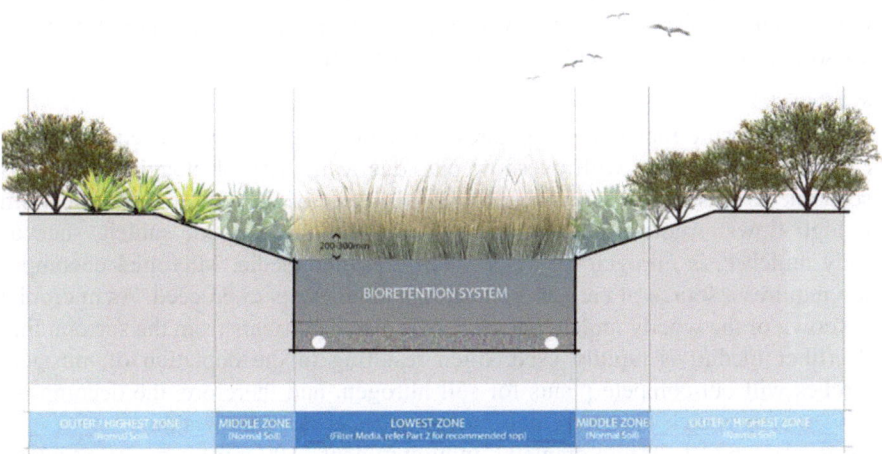

The key parameters to consider when designing with plants for biofiltration systems are:

Planting density
The overall planting density should be high. This will increase root density, maintain infiltration capacity and hence surface porosity. As a result, distribution of flows will be more even. Having dense planting will also increase evapotranspiration losses which reduce stormwater volume and frequency, and reduce weed competition. On the other hand, low density planting increases the likelihood of weed invasion and increases the subsequent maintenance costs associated with weed control.

Areas furthest from the inlet may not be ponded during small rain events in a large scale bioretention system. Plants selected for these areas may therefore need to be more drought resistant than those nearer to the inlet. On the contrary, plants near the inlet may be frequently inundated, and potentially buffeted by higher flow velocities. Therefore plants selected should be tolerant of these hydrologic impacts.

Range of species and types
A bioretention system with a range of plant species increases the success of the system as plants are able to "self-select" suitable establishment areas within the vegetated area—drought tolerant plants will gradually replace those plants that prefer wetter conditions (in areas furthest from the inlet).

Furthermore, bioretention systems with higher number of plant species and types have positive impacts on urban biodiversity compared to monoculture lawns. The presence of a bush canopy (mid-stratum) provides quality foraging and sheltering habitat for invertebrates that monoculture lawns cannot provide.

Where the landscape design includes mid-stratum, more shade tolerant species should be chosen for the groundcover layer. Trees and shrubbery should be managed so that the groundcover layer can still perform.

Use of mulch

The use of organic mulch such as hardwood chips is generally not recommended for bioretention systems with overflow pits, due to the risk of clogging. Mulch is susceptible to washout or will move to the perimeter of the system during a storm and high flows. Another reason for not recommending organic mulch, such as woody mulches, is nitrogen depletion from the filter media. Microbial decomposition requires a source of carbon (cellulose) and nutrients to proceed. As microbial breakdown of the woody mulch material takes place, nutrients from the surrounding soils (filter media) is rapidly used, often resulting in the depletion of nitrogen. Microbes will out-compete plants for soil nitrogen, and therefore, the decomposition of woody mulch may have detrimental impacts on plant health.

Stone mulch (10–20 mm diameter, minimum depth 100 mm) is preferred where there is a need to protect the soil from erosion or reduce the gradient of the batter slope (for safety reasons), whilst still maintaining the designed ponding volume.

A minimum depth of 50–100 mm gravel mulch is recommended to effectively prevent weeds from germinating and penetrating through the mulch layer. High planting densities should compensate for the reduced spread of plants caused by the stone or gravel mulch.

Safety consideration

The standard landscape design principles of public surveillance, exclusion of places of concealment and open visible areas apply to the planting design of bioretention basins. Regular clear sightlines and public safety should be provided between the roadway and footpaths or comply to the requirement of local authority.

Traffic sightlines

The standard rules of sightlines geometry apply. Planting designs should allow for visibility at pedestrian crossings, intersections, rest areas, medians and roundabouts.

Chapter 2
Selection of Plants that Demonstrated Nitrate Removal Characteristics

In a joint project between two agencies in Singapore, the National Parks Board and the National University of Singapore—Singapore Delft Water Alliance, more than 30 plants species were screened and tested to select those that are suitable for application as vegetation in bioretention systems. The research project investigated the remediation capacity of the plants and their associated rhizosphere microbial communities. Of the numerous stormwater pollutants, the phytoremediation study focused on nitrogen.

The experiments were set up and conducted from September 2010 through June 2011 at a site in Singapore, with an average day time temperature of 32 ± 1.87 °C and night time temperature of 25 ± 1.28 °C.

Experimental setup under a transparent pitched roof structure at a site in Singapore in 2010

© The Author(s) 2015

W.F. Hunt et al., *Plant Selection for Bioretention Systems and Stormwater Treatment Practices*, SpringerBriefs in Water Science and Technology, DOI 10.1007/978-981-287-245-6_2

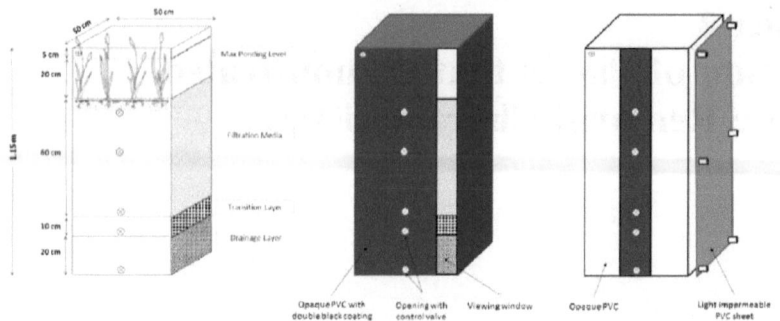

Schematic diagram of bioretention columns

The bioretention system comprised of three distinct layers: filtration layer, transition layer and drainage layer. The top filter media composed of 50 % Singapore's Approved Soil Mix (ASM) and 50 % medium to coarse sand. The transition media is made up of coarse sand and the bottom drainage media is made up of fine gravel. The filter media had a hydraulic conductivity of ca. 136 mm h^{-1}, which was compliant with the 50–200 mm h^{-1} range proposed in the bioretention design guidelines (PUB 2009).

Layer	Substrate	Particle size (mm)	Depth (mm)
Filter media	Mixture of 50 % ASM and 50 % medium to coarse sand	Varied	600
Transition layer	Coarse sand	0.7–1.0	100
Drainage layer	Fine gravel	1.0–5.0	200

To test the remediation capabilities of the plants for nutrients, irrigation water was chemically spiked to give a final concentration of 10 mg L^{-1} nitrate and 2 mg L^{-1} phosphate. These nutrient concentrations were above the levels commonly detected in urban stormwater runoffs, particularly in Singapore (Chua et al. 2009).

At the end of the experiment, the bioretention setups exhibited 100 % efficiency in removing phosphate. However, phosphate was also completely removed in the unplanted control, indicating that the remediation of phosphate was primarily attributable to the bioretention substrate.

Eleven plant species demonstrated high efficiency in nitrate uptake, removing more than 85 %. The plant information of the 24 species with more than 50 % nitrate uptake level is presented below.

Acalypha wilkesiana cultivar
Family name: Euphorbiaceae
Common names: Copperleaf, Joseph's Coat
Plant growth form: Shrub
Maximum height: 3.6 m
Maximum plant spread/crown width: 3 m
Origin: South Pacific Islands
Landscape uses: Hedge; border; mass planting;
container or aboveground planter; screen; Accent
Desirable plant features: Ornamental foliage
Light preference: Full sun
Water preference: Moderate Water
Percentage of Nitrate Removal: 79 %

Arundo donax var. versicolor
Family name: Poaceae (Gramineae)
Common names: Carrizo, Variegated Giant Reed,
Spanish Reed
Plant growth form: Shrub, Grass and Grass-like
Plant
Maximum height: 6.1 m
Origin: Asia
Landscape uses: Architectural accent/Screen
Desirable plant features: Ornamental Foliage
Light preference: Full Sun
Water preference: Moderate Water
Percentage of Nitrate Removal: 96 %

Bougainvillea **'Sakura Variegata'**
Family name: Nyctaginaceae
Common names: Bougainvillea
Plant growth form: Scandent shrub
Origin : South America
Landscape uses: Shrubs, Bushes, Groundcover,
Trellis, Accent
Desirable plant features: Ornamental Flowers
Light preference: Full Sun
Water preference: Little Water
Percentage of Nitrate Removal: 96 %

Bulbine frutescens (**L.**) **Willd.** *'Hallmark'*
Family name: Asphodelaceae
Common names: Orange Bulbine, Orange-stalked
Bulbine
Plant growth form: Shrub (Herbaceous; Creeper)
Maximum height: 0.3–0.6 m
Origin: Southern Africa
Landscape uses: Flowerbed/Border
Desirable plant features: Ornamental Flowers
Light preference: Full Sun
Water preference: Little Water
Percentage of Nitrate Removal: 68 %

Chrysopogon zizanioides (**L.**) **Roberty**
Family name: Poaceae (Gramineae)
Common names: Vetiver Grass
Plant growth form: Shrub, Grass and Grass-like Plant
Maximum height: 1.5 m
Origin: Tropical India
Landscape uses: Hedge, Screening, Accent Plant, Borders
Desirable plant features: Strong, deep roots for soil stabilisation
Light preference: Full Sun
Water preference: Moderate Water
Percentage of Nitrate Removal: 93 %

Codiaeum variegatum (**L.**) **Rumph. ex A.Juss.**
Synonyms: *Croton pictus*, *Croton variegatus*, *Codiaeum variegatum* var. *pictum*
Family name: Euphorbiaceae
Common names: Croton, Garden Croton, Puding, Joseph's Coat, Variegated Croton, Variegated Laurel, Joseph's Coat, Puding, Codiaeum
Plant growth form: Shrub
Maximum height: 3 m
Origin: South India, Ceylon and Malaya
Landscape uses: Indoor, General
Desirable plant features: Ornamental Foliage
Light preference: Full Sun
Water preference: Moderate Water
Percentage of Nitrate Removal: 67 %

Complaya trilobata
Family name: Asteraceae (Compositae)
Synonyms: *Wedelia trilobata, Wedelia carnosa, Silphium trilobatum, Thelechitonia trilobata, Sphagneticola trilobata* (L.) Pruski
Common names: Yellow Creeping Daisy, Singapore Daisy, Creeping Oxeye, Trailing Daisy, Rabbit's Paw
Plant growth form: Shrub
Maximum height: 0.15–0.70 m
Maximum plant spread/crown width: 2 m
Origin: Central America, South America
Desirable plant features: Ornamental flowers, ornamental foliage
Landscape uses: General, coastal, flowerbed/border, container planting
Light preference: Full sun, semi-shade
Water preference: Moderate water
Percentage of Nitrate Removal: 95 %

Cymbopogon citratus (**DC.**) **Stapf**
Synonyms: *Andropogon citriodorum, Andropogon citratus*
Family name: Poaceae (Gramineae)
Common names: Serai, Lemon Grass, West Indian Lemon Grass, Oil Grass, Fever Grass, Serai Makan, Sereh Makan
Plant growth form: Shrub, Grass and Grass-like Plant
Maximum height: 1.2–1.8 m
Maximum plant spread/crown width: 1.2 m
Origin: Southeast Asia
Landscape uses: General, Flowerbed/Border, Container Planting
Desirable plant features: Ornamental Foliage, Fragrant (Flowers: Time Independent; Foliage)
Light preference: Full Sun
Water preference: Moderate Water
Percentage of Nitrate Removal: 95 %

Dracaenaceae reflexa 'Song of India'
Family name: Dracaenaceae
Synonyms: *Pleomele reflexa* 'Variegata'
Common names: Song of India
Plant growth form: Shrub
Origin: South India, Ceylon
Desirable plant features: Ornamental Foliage
Landscape uses: General
Light preference: Full Sun, Semi-Shade
Water preference: Moderate Water
Percentage of Nitrate Removal: 64 %

Ficus microcarpa 'Golden'
Synonyms: Ficus nitida 'Golden'
Family name: Moraceae
Common names: Indian Laurel Fig
Plant growth form: Tree
Maximum height: 10 m
Origin: India, Malaysia
Landscape uses: House plantings, Hedge, Road-side Plantings, Bonsai
Desirable plant features: Ornamental Foliage
Light preference: Full Sun
Water preference: Moderate Water
Percentage of Nitrate Removal: 68 %

Galphimia glauca Cav.
Synonyms: Thryallis glauca
Family name: Malpighiaceae
Common names: Shower of Gold, Rain of Gold
Plant growth form: Shrub (Woody)
Maximum height: 1.0 m
Origin: Mexico to Guatemala
Landscape uses: Hedge/Screening, Groundcover,
Container planting
Desirable plant features: Ornamental Flowers
Light preference: Full Sun
Water preference: Little Water
Percentage of Nitrate Removal: 65 %

Ipomoea pes-caprae
Family name: Convolvulaceae
Synonyms: Ipomoea pes-caprae ssp. Brasiliensis,
Ipomoea biloba
Common names: Beach Morning Glory, Railroad
Vine, Sea Morning Glory, Tapak Kuda, Goat's
Foot
Plant growth form: Groundcover
Origin: Native to Singapore
Desirable plant features: Ornamental flowers,
ornamental foliage
Landscape uses: Coastal
Light preference: Full sun
Water preference: Lots of water
Flower Colour(s): Purple
Percentage of Nitrate Removal: 95 %

Leucophyllum frutescens (Berland.) I. M. Johnst.
Family name: Scrophulariaceae
Common names: Barometer Bush, Ash Plant, Cenizo, Texas Ranger, Texas Silver Leaf, Purple Sage
Plant growth form: Shrub (Woody)
Maximum height: 1.2–2.4 m
Maximum plant spread/crown width: 1.5 m
Origin: Rio Grande Valley (Texas, New Mexico), northern Mexico
Landscape uses: General, Coastal, Hedge/Screening, Flowerbed/Border, Container Planting
Desirable plant features: Ornamental Flowers, Ornamental Foliage, Ornamental Stems, Fragrant (Flowers: Day; [Remarks]: Peppery/spicy scent, reminiscent of carnations)
Light preference: Full Sun
Water preference: Little Water
Percentage of Nitrate Removal: 87 %

Loropetalum chinense (R. Br.) Oliv.
Family name: Hamamelidaceae
Common names: Chinese Loropetalum
Plant growth form: Shrub
Maximum height: 1.5 m
Origin: China
Landscape uses: Hedge, Garden Plant
Desirable plant features: Ornamental Flowers, Ornamental Foliage
Light preference: Full Sun
Water preference: Moderate Water
Percentage of Nitrate Removal: 71 %

Melastoma malabathricum L.
Family name: Melastomataceae
Common names: Common Sendudok, Singapore
Rhododendron, Indian Rhododendron, Sesenduk,
Malabar Gooseberry, Straits Rhododendron,
Sendudok, Senduduk
Plant growth form: Shrub (Herbaceous)
Maximum height: 1–3 m
Origin: Mauritius, Seychelles, Taiwan, Indian
subcontinent, Indo-China, Malesia, Australia
Landscape uses: Invasive/Potentially Invasive,
[Remarks] (Considered weedy. May form thick
thickets)
Desirable plant features: Ornamental Flowers
Light preference: Full Sun, Semi-Shade
Water preference: Moderate Water
Percentage of Nitrate Removal: 68 %

Nerium oleander 'Pink'
Family name: Apocynaceae
Common names: Oleander
Plant growth form: Shrub
Maximum height: 6–7 m
Origin: Unknown, probably South-west Asia
Landscape uses: Interiorscape/ Indoor Plant,
Hedges, Borders
Desirable plant features: Ornamental Flowers,
Fragrant
Light preference: Full Sun
Water preference: Moderate Water
Percentage of Nitrate Removal: 88 %

Ophiopogon jaburan
Family name: Convallariaceae
Common names: Lilyturf, Mondo-grass
Plant growth form: Groundcover
Origin: Japan
Landscape uses: Ornamental plants in gardening, houseplants
Desirable plant features: Ornamental foliage
Light preference: Full sun, semi-shade
Water preference: Moderate water
Percentage of Nitrate Removal: 77 %

Osmoxylon lineare (Merr.) Philipson
Synonyms: Boerlagiodendron lineare
Family name: Araliaceae
Common names: Green Aralia, Miagos Bush
Plant growth form: Shrub (Herbaceous)
Maximum height: 3 m
Origin: South-east Asia
Landscape uses: This species is used in Japanese-style gardens for its fine leaf texture
Desirable plant features: Ornamental Foliage
Light preference: Full Sun
Water preference: Moderate Water
Percentage of Nitrate Removal: 66 %

Pennisetum alopecuroides (L.) Spreng.
Synonyms: Panicum alopecuroides, Alopecurus hordeiformis, Pennisetum
hordeiforme, Pennisetum compressum, Pennisetum japonicum
Family name: Poaceae (Gramineae)
Common names: Chinese Fountain Grass, Swamp Foxtail Grass, Swamp Foxtail, Chinese Pennisetum
Plant growth form: Shrub (Herbaceous), Grass and Grass-like Plant
Maximum height: 0.6–1.5 m
Maximum plant spread/crown width: 0.5–0.6 m
Origin: India, China, Japan, Myanmar, Philippines, Indonesia, Australia
Landscape uses: General, Coastal, Flowerbed/Border, Container Planting
Desirable plant features: Ornamental Flowers
Light preference: Full Sun
Water preference: Lots of Water, Moderate Water
Percentage of Nitrate Removal: 93 %

Pennisetum x advena 'Rubrum'
Synonyms: Pennisetum setaceum 'Cupreum', Pennisetum setaceum 'Rubrum'
Family name: Poaceae (Gramineae)
Common names: Purple Fountain Grass, Red Fountain Grass, Rose Fountain Grass
Plant growth form: Shrub (Herbaceous; Creeper)
Maximum height: 1.2–1.5 m
Maximum plant spread/crown width: 0.5–1.2 m
Origin: Central Africa
Landscape uses: General, Coastal, Green Roof, Vertical Greenery/Green
Wall, Flowerbed/Border, Focal Plant, Container Planting
Desirable plant features: Ornamental Flowers, Ornamental Foliage
Light preference: Full Sun
Water preference: Moderate Water
Percentage of Nitrate Removal: 70 %

Phyllanthus myrtifolius Müll.Arg.
Family name: Euphorbiaceae
Common names: Ceylon Myrtle, Mousetail Plant
Plant growth form: Shrub
Maximum height: 0.5 m
Origin: India, Sri Lanka
Landscape uses: Flowerbed/Border, Bonsai
Desirable plant features: Ornamental Foliage
Light preference: Full Sun
Water preference: Moderate Water
Percentage of Nitrate Removal: 55 %

Sanchezia oblonga Ruiz & Pav.
Synonyms: Sanchezia speciosa, Sanchezia nobilis
Family name: Acanthaceae
Common names: Zebra Plant, Yellow Sanchezia,
Shrubby Whitevein, Gold Vein Plant
Plant growth form: Shrub (Herbaceous)
Maximum height: 1–3 m
Origin: Eucador, Northeastern Peru
Landscape uses: Hedge/Screening, Flowerbed/
Border, Interiorscape/
Indoor Plant, Container Planting
Desirable plant features: Ornamental Flowers,
Ornamental Foliage
Light preference: Full Sun, Semi-Shade
Water preference: Moderate Water
Percentage of Nitrate Removal: 87 %

Serissa japonica (Thunb.) Thunb.
Synonyms: Serissa foetida
Family name: Rubiaceae
Common names: Japanese Serissa, Snowrose,
Tree of a Thousand Stars, Japanese Boxthorn
Plant growth form: Shrub (Woody)
Maximum height: 0.45–0.60 m
Origin: India, China, Japan
Landscape uses: General, Coastal, Container
Planting, Bonsai
Desirable plant features: Ornamental Flowers,
Ornamental Foliage, Ornamental Stems, Fragrant
(Flowers: Time Independent; Foliage; Stems)
Light preference: Full Sun, Semi-Shade
Water preference: Moderate Water
Percentage of Nitrate Removal: 71 %

Scaevola taccada (Gaertn.) Roxb.
Synonyms: Scaevola sericea, Lobelia taccada,
Scaevola frutescens
Family name: Goodeniaceae
Common names: Ambung-ambung, Merambung,
Pelampung, Sea Lettuce, Sea Lettuce Tree
Plant growth form: Tree, Shrub (Herbaceous;
Creeper)
Maximum height: 10 m
Origin: S.E.A. to tropical Australia, Micronesia,
Melanesia, Hawaii and
Madagascar
Landscape uses: Coastal
Desirable plant features: Ornamental Foliage
Light preference: Full Sun
Water preference: Moderate Water
Percentage of Nitrate Removal: 88 %

Chapter 3
Inspection and Maintenance Guidelines

World-wide implementation of stormwater treatment practices is increasing. These are integral to Water Sensitive Urban Design (WSUD). In the United States, their implementation grew at a rapid rate in the late 1990s through the beginning of the millennium. For instance, in the state of North Carolina, nearly 20,000 stormwater treatment practices are in place. In Asia, Singapore is also seeing a rapid uptake of stormwater treatment practices, as part of its Active Beautiful and Clean (ABC) Waters' initiative.

Generally, much effort focused on design nuances for additional pollutant removal or better hydrologic performances. There is limited peer reviewed literature on maintenance issues, with a few notable exceptions (e.g., Asleson et al. 2009; Hunt et al. 2011; Wardynski and Hunt 2012).

The issue of maintenance, or a lack thereof, is generally recognized in the stormwater management industry worldwide. Practices that have been neglected do not mitigate runoff volumes or flows as well as intended (Brown and Hunt 2011).

3.1 Establishing Levels of Appearance

One principle that can be applied to determine maintenance requirements is establishing Levels of Appearance (LOA). LOA is a concept based on Levels of Service used in transportation structures such as roads and car parks, and this is equally applicable to water features.

LOA would specifically apply to aesthetic conditions. Two categories, Manicured and Rustic, have been established. The safety and hydraulic/hydrologic efficiency should not to be compromised and essentially would not vary between two LOAs, at all times.

The Manicured category includes highly visible practices, such as those in parks, in residential areas and along the roadway. Rustic practices are located in nature areas, or areas that mimic nature. They are expected to have a more natural appearance and therefore require less frequent upkeep.

© The Author(s) 2015
W.F. Hunt et al., *Plant Selection for Bioretention Systems and Stormwater Treatment Practices*, SpringerBriefs in Water Science and Technology, DOI 10.1007/978-981-287-245-6_3

The principal difference between the two levels of appearance is not so much the inspection and maintenance activities, but rather, the frequency of these tasks. As expected, Manicured LOA practices will necessarily be inspected more frequently (from daily to monthly), while Rustic LOA may only require visits quarterly to semi-annually.

Knowing the design intent of the practice can help establish an LOA. In general, a practice visible to the public will have a Manicured LOA, except when the setting makes park users expect a more Rustic look. On the other hand, a practice with little visibility by the public is more likely to have a maintenance regime aligned with a Rustic LOA.

For each practice, inspection and maintenance tasks are associated by their nominal frequencies. It should be noted that certain designs and systems may require more specific maintenance regimes. All Manicured LOA practices have a nominal long-term inspection frequency of once per month, but a more frequent inspection (such as weekly or daily) may be necessary, particularly immediately post-construction.

For cities with high rainfall distribution, the stormwater practices are designed to fill substantially with water on a frequent basis, and therefore require more frequent inspection.

It is also important to have in place a program to constantly train maintenance workers, in anticipation of turnover of personnel in maintenance positions.

The basis of maintenance for stormwater treatment practices is three-fold. Practices must be maintained sufficiently so that they are: (1) safe, (2) function hydraulically and meet water quality needs, and (3) aesthetic and litter free. The plant maintenance requirements will vary based on the LOA category.

The common practices, listed in general order of decreasing size are:

(1) Constructed Stormwater Wetlands and Wet Ponds,
(2) Sedimentation Basins,
(3) Bioretention (Landscape Infiltration) and Bio-Swales,
(4) Permeable Pavement,
(5) Swales,
(6) Infiltration Trenches, and
(7) Sand Filters.

In the subsequent paragraphs, a description of each practice and the respective inspection and maintenance needs is outlined.

3.1.1 Constructed Stormwater Wetlands and Wet Ponds

The distinction between ponds and constructed wetlands has blurred, as most ponds now incorporate wetland zones along their perimeter. Inspection for vegetative debris, litter/rubbish, and vandalism (in some cities) in both the outlet or inlet structure is critical. If the outlet structure clogs with debris, the pond's storage volume will decrease and plant survival becomes an issue.

A simple way to detect clogging is to determine whether the normal pool (water's surface) is abnormally high; this can often be accomplished by visual observation. Several other items should also be inspected:

(1) presence of odours indicating an acute accumulation of pollution,
(2) erosion in forebay/sedimentation zone or spillway
(3) animal damage, particularly from larger species like monitor lizards, and
(4) eutrophication, commonly identified by an abnormal amount of algae (possibly an algal mat) in the pond or wetland.

Certain maintenance activities can be performed during inspection. They include:

(1) skimming of floating or trapped litter/rubbish and vegetative debris and
(2) removal of invasive vegetative species

Another maintenance task is mowing the wet pond's banks and perimeter, as ponds often have manicured grass up to the water's edge. The frequency of mowing is dependent upon the required aesthetic appeal and safety issues pertaining to the wet pond and ranges from at least once per month (Manicured LOA) to as few as four times per year (Rustic LOA).

Of the monthly maintenance tasks for Manicured LOAs, the more important with respect to safeguarding the wetland or ponds' function are making sure the outlet is not clogged and verifying that there are no burrows or tunnels made by animals that artificially lower the water level inside the wet pond. If this maintenance is not performed, the hydraulic and water quality function of the wetland or wet pond will be compromised and a consequent high probability that the designed-for vegetation (vital to pollutant removal and aesthetics) will not survive.

Due to the design that includes an extensive coverage of emergent vegetation, stormwater wetlands run a higher risk of having their outlets clog than ponds.

Maintaining desirable plant species is also important for mosquito control. A few invasive species in the United States—particularly *Typha* spp. (cattails) and *Phragmites australis* (common reed), and *Salix nigra* (black willow) have been observed to form monocultures that may become habitats for mosquitoes (Greenway et al. 2003; Hunt et al. 2006).

Skimming rubbish from the pond will prevent future clogging of the drawdown outlet as well as improve the pond's aesthetics and eliminate a likely mosquito habitat.

Floating rubbish and debris can clog the outlet structure of a pond, harbour mosquitoes, and is unsightly

Certain inspection tasks can be performed much less frequently. These include:

(1) examining root intrusion into inlet or outlet pipes,
(2) assuring that the vegetated zones are, in fact, vegetated,
(3) determining the depth of sediment in the forebay,
(4) verifying dam health, and
(5) confirming the inlet rip rap apron is stable.

Forebay depth measurement is required annually

These inspection tasks may lead to the following maintenance tasks:

(1) clear inlet and outlet channels as well as internal drainage structures of roots,
(2) verify that emergency drawdown valves and appurtenances function, and
(3) sediment removal.

Cleanout is less frequent in stable catchments.

The frequency of tasks varies according to the configuration of the pond, watershed stability, and watershed composition. All of these maintenance tasks are essential to ensure both safety and the water quality and hydraulic function of wet ponds and wetlands.

The forebay should be excavated when sediment occupies 50 % or more of the forebay volume. Because the forebay in wetlands and wet ponds tend to be 1–1.5 m deep, a good rule of thumb is that if the average sediment level is within 0.5 m of the surface, the accumulated solids in the forebay should be excavated (dredged). For some cities, a second trigger mechanism for dredging could be whether any sediment or rubbish is visible at the water's surface (normal pool) or above in the forebay. The required frequency of forebay dredging varies from once per 6 months in developing, unstable catchments to being extremely rare (never or only once in the 20–30 year life of a wetland or wet pond) when sited in a stable catchment. If the forebay fills with sediment and is not removed, the remainder of the pond or wetland will begin to "silt in," essentially filling portions of the practice that are not designed for sediment capture. The main consequences of a silted-in stormwater practice are degraded pollutant removal capabilities and a degraded appearance. Excess sediment in the body of the wetland or wet pond will change the wetland plant population by creating areas elevated higher than as designed.

Constructed wetlands and wet ponds are designed to maintain a specific volume of water downstream of the forebay. This volume is to be preserved for stormwater runoff, thus, sediment accumulation in these areas is not desirable.

It is essential that wetland drawdown outlets are inspected for clogging. Clogged outlets may increase the water elevation inside the wetland for extended periods, potentially killing much of the desired vegetation. This will dramatically decrease the stormwater wetland's ability to remove certain dissolved pollutants, particularly some nitrogen species.

A clogged outlet can be easily fixed, but if it goes unnoticed due to a lack of inspection, the vegetative community of a stormwater wetland will be quite different from what the designer intended. This example demonstrates how long-term clogging eradicated nearly all the wetland plant species in what was intended to be a wetland-wet pond system.

Table 3.1 Common wetland and wet pond maintenance needs and drivers

Function impediment	Likely cause/s	Required maintenance
Wetland or pond water level elevated	Clogged outlet structure by litter or vegetation debris	Remove debris and litter from protective screen
	Recent heavy rain	Take no action. Let system drain
Sediment, debris and litter visible in forebay	Catchment produces a fair amount of debris, which has been caught	Excavate the forebay (backhoe digger most likely)
Water level too low	Leak in outlet structure or dam	Patch and repair outlet. Prevent large stemmed-vegetation from growing on banks
	Drought period	Take no action. Let it rain
Mosquito populations present in facility	Floating rubbish, debris, algae are harbouring mosquitoes	Remove algae, rubbish, and other debris through mechanical means
	Mass monocultures of vegetation are sheltering mosquitoes	See section on "removing unwanted species."
		Replace vegetation with that which attracts mosquito predators

The wetland and wet pond environment is a convenient place for some animals to live, especially monitor lizards. While these large lizards are skittish and avoid human contact, when they bed down in a wetland (or wetland-like fringes of a pond) they can damage vegetation. Maintenance personnel should also be wary of pythons and species of poisonous spiders that can live in wetlands and wet ponds.

A summary of the most likely maintenance needs and their triggers is presented in Table 3.1.

3.1.2 Sedimentation Basins

Sedimentation basins are related to wet ponds and wetlands, as often the latter two practices were initially designed to be sedimentation basins during construction. The purpose of sedimentation basins is to capture sediment generated by upstream construction. Thus, the main maintenance requirement for sedimentation basins is to remove collected sediment and other rubbish and debris. As sediment collects, the basin becomes more unsightly.

Most sedimentation basins take on a Rustic LOA, as they are associated with catchments under construction. However, if a sedimentation basin is located in an urban area that is frequented by people, its LOA may be raised to Manicured. By design, sedimentation basins are less attractive than other WSUL water features.

The ease of maintenance is dictated by the sedimentation basin design. To facilitate maintenance, designers should allow for access by excavators/diggers to all parts of the basin. If access to the basin is restricted, so that only a part of the basin can be reached by the bucket of excavators, maintenance cost in terms of personnel time increases dramatically.

A sedimentation basin with sediment collected.

The sediment depth should be checked frequently. A monthly regime is recommended. The designer should establish a threshold, which when met, triggers excavation of captured sediment. Thresholds could be:

(1) exposure of (or visible) sediment across a 10 % footprint of the sedimentation basins, or
(2) the average depth of sediment is within 0.5 m of the surface.

Setting stricter thresholds could be necessary if digger access is limited and sediment build-up needs to be minimal. If sediment from the basin is unable to be excavated using a digger, then small bobcats or crews with shovels and wheelbarrows are needed.

Clogged outlet structures and sediment accumulation are important concerns with sedimentation basins. Because they are usually constructed in catchments with active construction, the likelihood of rubbish accumulation inside the sedimentation basin is therefore high. If the outlet structure clogs, and water levels increase inside the basin, upstream flooding may occur. Like wet ponds, as sediment occupation increases, the volume available to capture stormwater diminishes. Inlet inspection and internal erosion are also concerns.

Common maintenance needs and triggers are listed in Table 3.2.

Table 3.2 Common sedimentation basin maintenance needs and triggers

Function impediment	Likely cause/s	Required maintenance
Excessive accumulation of sediment in basin	Catchment is unstable	Remove sediment and other gross solids using mechanical equipment
Water level in basin is elevated	Clogged outlet structure by litter or vegetation debris	Remove debris and litter from protective screen
	Recent heavy rain	Take no action. Let system drain

3.1.3 Bioretention Systems and Bio-swales

Bioretention systems, also termed biofiltration and rain gardens, are growing in popularity and usage in cities like Singapore. Locations for this practice include alongside roadways, in parks, adjacent to car parks, and in residential areas. Bioretention systems are expected to nearly always have a Manicured LOA, due to their locations that have easy public access.

Example of a bioretention cells located in a car park

Example of a bioretention cells located in a residential community

Example of a bioretention cells located in a park

Example of a bioretention cells located along a streetscape

Much of the maintenance needs of bioretention cells are aesthetic in nature. Considering that bioretention cells are sometimes used in lieu of standard medians in car parks or in place other landscape features, the maintenance requirements, while more costly per hectare treated than wet ponds and constructed stormwater wetlands, are actually very similar to standard landscape maintenance costs. Many maintenance tasks which are aesthetic also improve both safety and hydrologic and water quality performance.

Like any landscape feature, bioretention cells must be pruned and initially mulched, watered and limed. Some bioretention cells are grassed and require mowing. Because vegetation is an important investment and is essential to the aesthetic appeal of bioretention systems, vegetation needs to be established as quickly as possible. In some cases, the need for rapid establishment could require the bioretention system to be limed. This need may be established through conducting a soil test. Where soil alkalinity is a problem, ammonium sulfate may be added to the grounds. This may help establish vegetation within bioretention system.

Vegetation may need to be spot-fertilized to ensure rapid growth. If plant establishment occurs near a drought period, watering the plants every 2 to 3 days until a rainy period recommences is recommended. The frequency of these tasks varies based on the age of the bioretention cell. Newer bioretention cells require more frequent maintenance (weekly), while cells with well-established vegetation may only require monthly visits.

Various bioretention maintenance activities include: annual to semi-annual pruning (top), and an initial fertilization to ensure plant survival (bottom).

In temperate climates, various bioretention maintenance activities include removal of biological films which cause the bioretention cell to clog; this is needed every 2 to 3 years. Fully vegetated bioretention systems in the tropics may not need this type of maintenance frequently.

Bioretention outlets are prone to clogging. It is imperative that outlets are checked on a monthly basis to ensure they are free of litter and debris. This is particularly important if a grate is used on the highflow bypass structure. Nearly every bioretention cell has an overflow, so if it cannot be found, the location should be checked using the engineering design plans.

Should vegetation and rubbish continue to collect on the grate, water may spill into the car park.

Some bioretention cells in temperate climates have been found to clog due to sediment accumulation (Brown and Hunt 2011). Dense and uniform root depth helps to maintain good infiltration throughout the cell, as has been shown in Australia (Bratieres et al. 2008; Lucas and Greenway 2008).

When bioretention cells are surrounded by landscapes under active construction, they are at high risk of clogging by sediment. In these bioretention cells, it is important to inspect the bowl for water collection and to identify any wet spots in the media, particularly if it has not rained in at least 2 days. If surface clogging is observed, the top layer (25–100 mm) of fill soil needs to be removed. Because clogging occurs most frequently at the top of the soil column, the bioretention cell rarely needs to be completely excavated.

While not included as part of bioretention maintenance, protecting the perimeter of the bioretention cell from construction-generated sediment has a dramatic impact on the severity of clogging and the type of maintenance required resuscitating the bioretention cell's infiltration rate.

To prevent premature clogging of bioretention cells due to sediment accumulation, designers are recommended to incorporate pre-treatment. The most commonly used pre-treatment in temperate climates in descending order are:

(1) gravel verge (thin strip) with sod surrounding the perimeter,
(2) grass swale, and
(3) forebays.

Installing sod on the down slopes of the pavement provides an immediate layer of pre-treatment before runoff enters the bioretention cell proper. The sod is part of a grassed filter strip. The minimum width required for the sod filter strip is 1 m, with 1.5 m recommended. In addition to trapping pollutants before they reach the

bioretention cell media, the sod immediately stabilizes the perimeter of the bioretention cell, preventing "internal" erosion from occurring. In the tropics, cowgrass
(*Axonopus compressus*) and Nabhali (*Cyanotis cristata*) most are resilient turf
grasses often considered. The larger the grass filter strip, the more important mowing
becomes as part of its maintenance regime. The grassed filter strip functions best
when it is not level with the pavement. A 50–75 mm drop from the pavement edge to
the filter strip is recommended. This prevents preferential flow through the filter strip
and reduces the risk of rill erosion.

A gravel and sod verge (top) or solely a grassed filter strip (bottom) can help limit the amount of
maintenance performed on the filter media in the main part of the cell.

A 70-mm difference in elevation between pavement edge and turf grass reduces the risk of rill erosion.

Like those of wetlands and wet ponds, bioretention forebays capture larger sediment particles and litter/debris. They should be checked regularly and noticeable accumulations of sediment and litter should be removed as needed. The frequency of this maintenance task is associated with the cell's LOA, but is expected to range from a few times a year to once every 3–5 years.

Another practice that is becoming common in cities is the bio-swale. Bio-swales provide similar pollutant treatment mechanisms as bioretention cells, and they also filter water through a permeable media. The two principal differences between bioswales and bioretention are

(1)   bioretention is designed to pond water, while bio-swales are not and
(2)   bio-swales convey water, while bioretention cells pond water.

Examples of bioswales along a promenade (top) and adjoining a car park (bottom)

Outlet structures in bio-swales need not become water collection points, and are therefore nearly flush with ground level. Checking outlet structures will be paramount to bio-swale success.

Like bioretention cells, most bio-swales are expected to be Manicured LOA, thereby requiring weekly to monthly inspection and simple maintenance. All of the maintenance tasks reviewed for bioretention is applicable to bio-swales.

Table 3.3 summarises the most likely maintenance needs and their triggers.

Table 3.3 Common maintenance needs and triggers for bioretention and bio-swales

Function impediment	Likely cause/s	Required maintenance
Surface water present after at least 1 day since rainfall	Sediment accumulation in media is restricting flow	Remove collected sediment, typically to a depth of 75–150 mm
	Media itself is not sufficiently permeable	Test permeability of media. Media may need to be replaced
Vegetation is dying	Cell is too wet, and not draining sufficiently fast	See maintenance tasks suggested above
		Consider replanting with wetland vegetation
	Cell is too dry, and not receiving enough runoff	Adjust catchment to increase runoff volumes (if possible)
		See if outlet device has leaks, leading to prematurely drainage
		Consider replanting with more drought-tolerant vegetation
	Soil media is too sterile	Take soil test. Add alkalinity if needed, spot fertilize vegetation
Mosquitoes are present in bioretention cell or bio-swale	Water is present for too long at surface	See related maintenance tasks
	Litter is harboring mosquitoes	Remove rubbish
Water is overflowing the bioretention cell or bio-swale regularly	Outlet structure is too close to the soil surface	Consider raising outlet structure, but designers MUST be contacted
	Outlet structure is clogged, causing overflow in undesired locations	Remove vegetation or debris and rubbish from overflow grate

3.1.4 Sand Filters

The sand filter comprises two main chambers: a sedimentation chamber, where gross solids and some sediment collect, and a sand chamber which filters some pollutants which remain in the runoff. Sand filter maintenance is relatively simple, but somewhat laborious and expensive. Both chambers require inspection and maintenance on regular intervals. Because sand chambers are usually hidden from the public, they are likely to require a Rustic LOA.

On a monthly to quarterly basis, the sand chamber needs to be checked to see if the top layer of sand has become clogged with fine particles, which limits the infiltration rate through the soil. This clogging is identifiable by a thin (few millimetres in thickness) black or dark brown/gray layer resting on the sand surface. This is referred to as the smutzdecke, German for "dirty floor". If this layer of smutzdecke is observed at a given site, it should be broken up using a hard garden rake. Typically only the top 25–50 m forms a crust, therefore the depth the rake needs to penetrate is not substantial. The broken up smutzdecke does not need to be removed regularly, because it contains microbes that facilitate the breakdown of petroleum products. The frequency of this maintenance task varies, but is expected to occur between quarterly to annually.

Gross solid accumulation in the sedimentation chamber

A smutzdecke (or polluted fines layer) in the sand chamber.

However, when the sedimentation chamber needs to emptied of the accumulated debris, which occurs much less frequently than breaking up the smutzdecke, it is advisable to then excavate the top layer of sand (perhaps to a depth of 75–100 mm) and transport this material away with the spoils from the sedimentation basin.

When the material is taken from the sand chamber it must be replaced. The type of sand needed to "top off" the sand chamber should be specified in the design plans. If the design plans are not available, a clean sample of the existing sand media can be found at a depth of 20–30 cm in the current sand media chamber. Taking a sample at this depth and analyzing it will allow the proper sand media needed to "top off" the sand chamber to be identified. It is possible the sample will need to be send to an engineering laboratory.

The depth of gross solids in the sedimentation chamber should be inspected each site visit; one can use a shovel or a crowbar. If the sedimentation chamber is at least one-half full of detritus and trash, then the accumulated debris should be removed. Additionally, if a stench arises from the sedimentation chamber, it is likely a good time to remove the accumulated solids, even if the depth of solids is not yet 50 % of capacity. Removal of debris can be accomplished with a vac-truck or a small excavator.

The process for permanently disposing sand filter pollutant material is currently under debate. In some countries, the disposal practice is to deposit the accumulated pollutant in a lined landfill. In future years, it may be necessary to have the polluted media tested to determine the toxicity of its contents before landfill disposal is allowed.

Shoveling accumulated gross solids and sediment into the bucket of a backhoe.

Table 3.4 Common sand filter maintenance needs and triggers for sand filters

Function impediment	Likely cause/s	Required maintenance
Chambers omit stench	Accumulation in sedimentation chamber of trash and debris is decaying	Remove sediment and other gross solids using mechanical equipment
		Simultaneously remove 75–100 mm from top of sand chamber
Water level in sand chamber is elevated	*Smutzdecke* has formed and is restricting flow through media	Break up *smutzdecke* to a depth of 50 mm with hard garden rake
	Sand media is clogged by sediment, litter, and debris	Remove debris and litter from top 75 to 150 mm of sand chamber

Sand filters are often accessible to vehicle traffic and as a result have heavy grates overlaying them. Removing and replacing metal grates is an additional expenditure of time and effort. Another unique sand filter maintenance need is to remove vegetation and/or trash blocking the path of water from the sedimentation chamber to the sand chamber, which impedes water passage. If a sand filter still has collected a pond of water for more than 12 h after a rain event, it is likely clogged.

Table 3.4 summarise the likely sand filter maintenance needs and triggers.

3.1.5 Swales

Swales are used primarily to convey water, but also provide treatment of solids and pollutants associated with sediment. Swales are related to bio-swales, in that they both convey water, but bio-swales have a specialized fill media and tend to be vegetated with a larger variety of species. Most swales are predominantly turfgrass.

Examples of swales in Singapore

Swales are maintained similarly to their surrounding landscape. Mowing and weeding at regular intervals ensures a clean appearance. Care must be taken to avoid mowing when the swale is wet, otherwise rutting may occur. Also, mower blades must be sufficiently high so that the turfgrass is not scalped or cut too low. It is recommended to set the height of the mower at least 70 mm high, with 100 mm preferable, for turfgrass health.

Because swales convey runoff, they also transport litter and other solids. Depending upon the surrounding land use, it may be necessary to remove litter weekly.

Rubbish collecting in the trough of a swale, with a modest amount of erosion.

Occasionally, the swale will receive flows exceeding their capacity. This results in either overtopping of the swale's banks or erosion within the swale. Internal swale erosion is most likely apparent after a very intense rain event and should be repaired shortly after discovered. Eroding swales can "migrate" causing a more expensive repair to be needed with time. The simplest fix is to reline the toe of the swale with rocks or perhaps use a turf reinforcement mat that allows grass growth but provides erosion-resistant structure.

Turf Reinforcement Mats are an alternative to rock-line swales.

When swales are constructed in flat topography, or they are built as small cascades, they run the risk of pooling water. Water pooling in swales facilitates mosquito survival. While wetlands and wet ponds also have standing water, they also provide habitat and attract mosquito predators. Bio-swales and bioretention cells both drain water through a media, reducing the likelihood of mosquito presence. Swales, however, neither attract mosquito predators with their vegetation, nor do they have a mechanism to dewater.

Water collected in a swale can become a mosquito habitat.

There are two basic solutions to the small pools within swales:

(1) to drain these pools, by changing the plan form of the swale (increasing its slope) or
(2) ensure pools are identified and provide mosquito treatment within them.

Biocides are available in many parts of the world; they are called "mosquito dunks" in the USA.

Sometimes swales will intersect high water tables or have water backed into them from nearby waterways. There is very little that can be done in this situation, as the turf will die. Sometimes these wetter swales can be transformed into linear wetlands. Mosquito issues will always be of concern here, particularly in the rainy season. Depending on the water depth, lining the bottom with larger rocks might be a feasible and more attractive.

Maintenance of swales with collected water should focus on mosquito prevention.

A summary of the most common swale maintenance needs and triggers is presented in Table 3.5.

Table 3.5 Common maintenance needs and triggers for swales

Function impediment	Likely cause/s	Required maintenance
Sediment is visible in swale	The swale is eroding	Line swale with flow-resistant rocks or turf reinforcement liner
	Sediment from catchment is accumulating in swale	Stablise the catchment
Mosquitoes are present in swale	Swale is not completely dewatering. Some pools remain	If some relief, consider breaching small pools
		If no relief, consider mosquito dunks
	Litter and debris are providing shelter	Remove litter and debris
Turf grass appears unhealthy	Mowing too low during times of drought	Reviewed in "mowing and turf grass" section
	Infertile soil	Take soil test. Potentially add lime and/or spot fertilize

3.1.6 Infiltration Trenches

Infiltration trenches resemble swales and bio-swales. However, they are not designed to have vegetated cover. As its name implies, they are specifically intended to infiltrate runoff. Infiltration trenches may have either a Manicured or Rustic LOA, depending upon their location.

The primary maintenance concern for infiltration trenches is sediment collection. The gravel trench will clog with substantial sediment accumulation. The infiltration trench's catchment should be inspected regularly. Sediment may migrate deeper into the gravel column, but not be visible at the gravel surface, so digging into trench, perhaps to a depth of 150 mm, is important to detect long-term clogging sediment.

While not evident at the surface, sediment can migrate into the gravel column of the infiltration trench. This can be inspected for by digging several cm into the gravel.

Visiting infiltration trenches during storm events is a good way to determine if the trench is working. Infiltration rates of non-clogged gravel can easily exceed 1,000 mm/h, which exceeds precipitation + runoff rates into the infiltration trench. If during a rain event, runoff appears to flow right over the infiltration trench, it is likely clogged.

If the trench is inspected outside of a rain event, other signs of clogging besides exposed sediment include moss and other vegetation growing on the surface, as well as collected water. The repair may be costly, as the gravel trench will need to be removed, accumulated sediment excavated from the exposed trench, and clean gravel replaced in the newly-formed cavity.

Moss, other vegetation and water at the surface of an infiltration trench are signs that the practice
is clogged and not infiltrating.

When infiltration trenches are constructed in flat landscapes or other locations
where the water table emerges at or is near the surface, the infiltration trench will
not work (lack of infiltration). There is nothing that can really be repaired by
maintenance personnel; the issue is a design and site-specific problem.

Table 3.6 Common infiltration trench maintenance needs and triggers for infiltration trenches

Function impediment	Likely cause/s	Required maintenance
Water is ponding at surface of trench	Gravel fill is clogged by sediment, litter, and debris	Stabilize catchment to eliminate source of sediment
		Excavate and replace gravel fill layer, as it is likely clogged top to bottom
	Trench intersects high water table	Nothing can be done without re-design. Consider creating linear wetland?
Moss growth is apparent on gravel surface	Same as those for "water is ponding at surface of trench"	Same as those for "water is ponding at surface of trench"
Mosquito populations present in Infiltration trench	Clogging or high water table intersection	Consider adding mosquito dunks
		Consider converting to more natural and mosquito resistant practice (like linear wetland) that attracts mosquito predators

A summary of common infiltration trench maintenance and triggers is found in Table 3.6.

3.1.7 Permeable Pavements

Permeable pavements are widely adopted in North America and Europe. When designed and installed correctly, permeable pavements effectively reduce runoff volumes and pollutant loads. Because permeable pavement is most likely used in car parks, it is nearly always going to have a Manicured LOA.

Permeable pavement installations in Singapore (top) and Illinois, USA (bottom).

Permeable pavement allows water to pass through the surface and be stored in an underlying gravel storage layer. Water then drains from the gravel storage zone or infiltrates into the in situ soil.

Functionally, permeable pavement must remain permeable. As the surface of permeable pavement clogs, the pavement will begin to generate runoff. Main causes of clogging are vegetation/detritus and sediment accumulation in the permeable gaps of the pavement. Therefore, the principal maintenance tasks associated with permeable pavement focus on prevention of complete surface clogging.

Examples of runoff occurring from "permeable" pavement

Permeable pavement should be inspected during heavy rain events. Because by design, permeable pavement is expected to have infiltration rates exceeding 1,000 mm/h. Even the most intense rain should fail to generate runoff. Anywhere that runoff is observed to occur is therefore clogged.

There are areas most susceptible to clogging:

(1) landscape-hardscape interface (particularly including pavement located under tree canopies),
(2) impermeable pavement-permeable pavement interface, and
(3) along the paths of "dirty" vehicles (such as rubbish collecting trucks).

Thus, these are the prime locations where runoff generation may occur and these should be inspected most frequently. If the surface of the permeable pavement has either turned dark in colour or has vegetation growing in it, then the pavement is likely clogged. Either is a sign of sediment and/or organics collection at the surface.

Table 3.7 Comparison of street sweepers for permeable pavement

Street sweeper type	Degree of clogging	Penetration of smutzdecke	Runoff generation
Regenerative air	Minor/early stages	1–5 mm	Sometimes
Vacuum	Major/long-term neglect	Up to 40 mm	Always

Clogged surface, or smutzdecke remnant at front of paver.

If clogging is observed, the pavement should be maintained. In many parts of the world, this is done with a street sweeper. The general type of sweeper used depends upon the permeable pavement type and the severity of clogging (Table 3.7).

If a clogged permeable pavement application is unreachable by a street sweeper or vacuum truck, the surface can be painstakingly unclogged using a portable industrial strength vacuum cleaner, referred to as a "shop vac" in North America. To successfully clean permeable pavement using a "shop vac," suction must be hand applied to every joint that is clogged. Often multiple passes are needed.

Using an industrial strength vacuum cleaner to unclog permeable pavement.

Ideally, vegetation and detritus that could potentially clog permeable pavement is removed before it does so. This may be accomplished by air blowing detritus off the pavement. This is clearly preventative maintenance. The frequency at which air blowing should occur is dependent upon rates of leaf litter collection on the surface of permeable pavement.

Air blowing detritus from the pavement surface before it forms a smutzdecke is a first line of defence.

Table 3.8 Common permeable pavement maintenance needs and triggers

Function impediment	Likely cause/s	Required maintenance
Runoff is observed from the pavement	Sediment or vegetative debris is clogging the surface of the pavement	If sediment, try to restrict the run-on of sediment, by stabilising the catchment
		Run a street sweeper (see Table 3.7) over the clogged portions of the pavement
The surface of the pavement has turned dark	Organic debris collection. This is a likely spot for runoff production	Run street-sweeper
		Consider increasing frequency of air blowing vegetation from surface of pavement
Discoloring and stains are present on surface	Vehicle leaks	Apply biodegradable detergent and water blast the surface
		Take no action. Accept that stains occur on pavement

Permeable pavement, like any pavement, is at risk to stain due to vehicle leaks. Assuming a Manicured LOA is needed, these stains need to be removed. Biodegradable detergents can be applied to oil and grease stains and then water blasted from the pavement surface. This has anecdotally removed 85–90 % of the stain, with the remainder resembling something like a water mark. If a Rustic LOA is acceptable, and provided the permeable pavement has a darker hue, it may be reasonable to forego stain removal.

Table 3.8 highlights the most common permeable pavement maintenance needs and triggers.

3.2 Mowing and Turf Grass

Nearly every vegetated practice can be partially comprised of turf grass. Swales are turf grassed-lined long their batter slopes and, if small enough, their "channel" bottom. Bioretention employs turf grass as a short filter strip that water passes through from a paved surface. The perimeter of ponds and wetlands are often ringed by turf as part of their desired appearance. This turf grass is often intended to be a final gross solids filter for runoff prior to entering the pond or wetland.

As such, after keeping practices litter free, mowing turf grass is the most unilateral and common maintenance practice. In general, the higher the shoot (or blade) of the grass is, the deeper the root. Deeper rooted grasses are better able to tolerate weather extremes, namely droughts. Ideally then, mowing regimens and recommendations should reflect the desire for taller, but still aesthetically pleasing, turf grasses.

Table 3.9 Guide to mowing height and frequency

Location	Mowing height	Frequency	Note
General landscape (not in WSUL practice)	40 mm	Nominally each 2 weeks	Frequency of every 3 weeks if dry. Every week if wet
Batter slope on swales	100 mm	Nominally each 2 weeks	Frequency of every 3 weeks if dry. Every week if wet
Filter strip perimeter along Bioretention	75–100 mm	Nominally each 2 weeks	Frequency of every 3 weeks if dry. Every week if wet
Filter strip perimeter along wetlands and wet ponds	100–120 mm	2–4 weeks	Perhaps less frequent if LOA = Rustic

Cow grass, a commonly used in the tropics, forms a very dense mat. It also has a dwarf variety. Table 3.9 establishes targeted mowing cut heights for cow grass with respect to location in the landscape. Mowing or cutting is site- and technique-specific.

It should be noted that, particularly in bioretention and swales, the underlying soils are prone to be wetter than those in surrounding landscapes. This is due to runoff passing over this turf. Because the underlying soils tend to be wetter, mower rutting is more likely. Maintenance personnel will need to be mindful of rutting and either use different equipment or wait for a slightly drier condition before mowing (if possible).

3.3 Removing Unwanted Invasive Species

Wet ponds with aquatic shelves, stormwater wetlands, poorly draining bioretention cells, and even swales can become overgrown with invasive species. A few most common invasive species are *Eichhomia crassipes* and *Hydrilla verticillata*. Cattails (*Typha angustifolia* spp.). Cattails are tolerant of a variety of environmental conditions and can sequester pollutants (Wang et al. 1993). From a function standpoint, cattails may be considered a good species of plant to have in any of stormwater treatment practices. However, a practice that is overgrown with cattails is not a diverse ecosystem. Ecosystem diversity is critical for mosquito control. Rafts of cattails provide a safe environment for mosquito larvae to mature to adulthood.

Removing cattails can be tricky. Cattails grow by rhizomes and also spread by seed. If a piece of rhizome is left in the soil after cattail removal, the stand will re-establish. One means of mass cattail removal is by way of a digger. This is recommended when a wet pond or wetland is completely overgrown by cattails.

If a stormwater practice, particularly a wetland or wet pond, has a diversity of vegetation, but cattails are beginning to colonize, broadcast spraying of herbicides must be avoided, as broadcast spraying will kill not only the cattails, but other desirable vegetation as well.

The frequency of cattail removal somewhat varies. Factors influencing the need to "wipe" cattails are the density at which the wetland or wet pond is planted with desirable species and the maturity of the practice. It has been observed in temperate and sub-tropical climates that during the first 1–2 years, cattails should be removed twice annually. As the wetland matures and desirable species begin to firmly establish, the maintenance frequency is reduced to once annually at the most. The amount of time needed to remove unwanted vegetation varies, but a well-maintained mature wetland will likely require between 20 and 30 min per 0.10 ha of impacted wetland (or wetland shelf within a pond) per visit.

Simple plant removal, like weeding, may sufficiently control invasive and other undesirable species if performed at a very high frequency—such as daily visits. It is critical that maintenance personnel be trained in the appearance of undesirable species early in their life cycle. This technique of weed control is labour intensive, particularly initially. As the practice matures, the frequency decreases.

Daily weeding is an effective, but time intensive, way to control unwanted vegetation.

References

Asleson, B.C., Nestingen, R.S., Gulliver, J.S., Hozalski, R.M., Nieber, J.L.: Performance assessment of rain gardens. J. Am. Water Resour. Assoc. **45**(4), 1019–1031 (2009)

Bratieres, K., Fletcher, T.D., Deletic, A., Zinger, Y.: Nutrient and sediment removal by stormwater biofilters: a large-scale design optimisation study. Water Res. **42**(14), 3930–3940 (2008)

Brown, R.A., Hunt, W.F.: Impacts of media depth on effluent water quality and hydrologic performance of undersized bioretention cells. J. Irr. Drain. Eng. **137**(3), 132–143 (2011)

Greenway, M., Dale, P., Chapman, H.: An assessment of mosquito breeding and control in four surface flow wetlands in tropical-subtropical Australia. Water Sci. Technol. **48**(5), 249–256 (2003)

Hunt, W.F., Apperson, C.S., Kennedy, S.G., Harrison, B.A., Lord, W.G.: Occurrence and relative abundance of mosquitoes in stormwater retention facilities in North Carolina, USA. Water Sci. Technol. **54**(6–7), 315–321 (2006)

Hunt, W.F., Greenway, M., Moore, T.C., Brown, R.A., Kennedy, S.G., Line, D.E., Lord, W.G.: Constructed stormwater wetland installation and maintenance: are we getting it right? J. Irrig. Drain. Eng. **137**(8), 469–474 (2011)

Lucas, W.C., Greenway, M.: Nutrient retention in vegetated and nonvegetated bioretention mesocosms. J. Irrig. Drain. Eng. **134**(5), 613–623 (2008)

Wang, S.C., Jurik, T.W., Vandervalk, A.G.: Effects of sediment load on various stages in the life and death of cattail (Typha x glauca). Wetlands **14**(3), 166–173 (1994)

Wardynski, B.J., Hunt, W.F.: Are Bioretention cells being installed per design standards in North Carolina? A field assessment. J. Environ. Eng. **138**(12), 1217–1220 (2012)

Yong, J.W.H., Yok, T.P., Hassan, N.H., Ngin, T.S.: A Selection of Plants for Greening of Waterways and Waterbodies in the Tropics. National Parks Board, Nanyang Technological University and Public Utilities Board, Singapore (2010)

© The Author(s) 2015
W.F. Hunt et al., *Plant Selection for Bioretention Systems and Stormwater Treatment Practices*, SpringerBriefs in Water Science and Technology, DOI 10.1007/978-981-287-245-6